Sounds of Animals at Night

EDWARD R. RICCIUTI

Sounds
of Animals at
Night

by Edward R. Ricciuti

Illustrated with Photographs

Harper & Row, Publishers

New York, Hagerstown, San Francisco, London

Sounds of Animals at Night
Copyright © 1977 by Edward Raphael Ricciuti

FIRST EDITION

Library of Congress Cataloging in Publication Data

Ricciuti, Edward R.
    Sounds of animals at night.

    Bibliography: p.
    Includes index.
    1.   Nocturnal animals—Juvenile literature.
2.   Animal sounds—Juvenile literature.   I. Title.
QL755.5.R5   591.5′9   76–3843
ISBN 0–06–024980–3
ISBN 0–06–024981–1 lib. bdg.

To Tina, who thought of it

Contents

1 Noises in the Night

Many different kinds of animals are awake and active at night. As they go about their business, they usually stay out of sight. But some of them make noises that we can hear. They scuffle and scratch. Whistle and buzz. Hoot, howl, and screech. Squeak, skirl, and croak.

Noises of night animals often sound weird, but each has a meaning. It may be a danger signal, a warning cry, or a mating call. It may mean that an animal is feeding or is in trouble. Or it may be just an accidental noise, like the scratching of a raccoon's claws as it climbs a tree. Noises in the night can tell us what sorts of animals are near, and what they are doing. The calls of night animals can even be clues to what kinds of weather to expect. All we need is an understanding of the sounds we hear.

When we hear sound, we are really hearing vibration. For

example, think what happens when a guitar string is plucked. Each time the string vibrates, it pushes out against the molecules of air around it. This sets up a series of waves in the air. You might picture such sound waves by imagining the ripples that fan out from a rock hitting the water.

When sound waves strike the eardrum, it vibrates. In the ear the vibrations are changed to nerve messages which speed to the brain. There, the nerve messages are heard as sound. Its pitch depends on the rate of vibration. The faster the vibration is, the higher the pitch.

The sound of the human voice is made by the vibration of two elastic vocal cords in the throat. The cords are stretched over a frame of cartilage, called the voice box, or larynx. They are made to vibrate by air from the lungs.

The larynx of other mammals is similar to the human voice box. Frogs and toads also have a larynx. Two elastic lips take the place of vocal cords in the voice box of these amphibians. Birds have a voice box called a syrinx. Instead of vocal chords, the syrinx has elastic membranes which vibrate under the force of air from the lungs.

For a sound to be considered a true voice, it must come from either a larynx or a syrinx. The larynx is found only in mammals and amphibians. Only birds have a syrinx. So only these three groups of animals have true voices.

Many other animals make sounds, but not with voices. Some reptiles, for example, hiss or grunt by letting out a blast of air.

Many insects "sing" by rubbing one part of their body against another. But these sounds are not really voices.

Many of the calls made by animals are very difficult to describe with words. We often try to mimic animal calls with words that sound like them, such as *chirp* and *hoot.* Or else we liken them to more familiar sounds, such as the creaking of a gate or the piping of a flute. The descriptions are not always perfect, but usually they are the best we can do.

This book will describe some of the animal sounds that you are likely to hear at night. It will explain how they are made and why. With the help of what follows, you should be able to identify many night noises that once may have seemed strange.

2 The Frogs and Toads

Sleigh bells, creaky gates, and jug o' rum

The mating calls of frogs and toads are among the strangest night sounds. They often mystify people. No wonder. Sometimes it is very hard to imagine that amphibians can make such sounds. They may seem more like the calls of birds or insects, or maybe even like noises made by nonliving things. Some are sweet and musical; others are harsh; still others are comical. But they all have the same important meaning: breeding season has arrived. The males are ready to mate with the females.

Only the male frogs and toads make mating calls. The caller seals in a gulp of air by closing its nostrils and mouth. The air is pumped back and forth between the lungs and throat, over the voice box. The sound made by the voice box builds up in the throat, which expands into a vocal sac. Some small species have very large vocal sacs—almost as large as they are. Other species have a pair of smaller sacs, one on either side of the head. The vocal sac permits even a small frog or toad to call loudly.

Each kind of frog and toad has its own special call. A female of the same kind recognizes it. It brings her to the male so they can mate.

Each kind also has its own particular breeding season. For most frogs and toads, it is during the period between late winter and the following autumn. The exact time depends on the species and, often, where it lives.

Once the breeding season begins, the males call for hours at a stretch. Some call in the day as well as the night. After dark, however, the frog-and-toad "chorus" is loudest.

The calls in the chorus change as the breeding seasons for different species begin and end. By knowing which species are calling, it is possible to determine the time of year without using a calendar.

Over much of North America the first caller of the year is the wood frog. This frog begins to call as soon as the ice starts to melt. It can be heard as early as January in the southeastern United States. By late February or early March the wood frog is calling in New England.

The call of a lone wood frog sounds like the creaking of a rusty gate. Several wood frogs calling together sound like the quacking of a flock of ducks. In fact, wood frogs often fool people into thinking they have heard ducks, but the ducks never appear.

As the breeding season begins for the wood frog, swarms of

wood frog

males gather in quiet swamps and small ponds. They arrive suddenly and silently. But then the quiet of the woods is shattered as they explode into a furious burst of calling. It sounds like the words *quacketyquacketyquackety* said very fast.

A group of wood frogs can make a lot of noise. But the sound does not travel very far, only about a hundred yards.

Sometimes the wood frogs begin calling, then stop as suddenly as they started. Then they begin again, call for a few minutes, and stop once more. They may do this several times before they really begin their chorus.

If there is a sudden freeze, the wood frogs may stop calling for several days. They swim to the bottom of ponds and creep into the dead leaves and mud. They do not stir until the weather warms up again. Then they begin to call as if nothing has happened.

The wood frogs call while floating at the surface, with legs outspread. Even a small pond may have hundreds of wood frogs dotting its water. Every so often one of the frogs may kick through the water, straight at another one. Sometimes the two frogs collide. Such behavior is probably a contest for the right to call from a particular patch of water.

Drawn by the calling, the female frogs come out of the woods. They are slightly larger than the males, which are two to three inches long. The males are brownish, while the females are reddish-brown. Both have a dark "robber's mask" over their eyes.

Once a female is in the water, several males may swim after her. The first male to reach her tries to climb onto her back. If he manages to get on, he holds tightly. He does not fall off even when the female swims. As the female lays her eggs in the water, the male fertilizes them. Most other frogs and toads also mate this way.

Within a month after the first wood frogs have called, the breeding season ends. The frogs leave the water and fan out through the woods. In the western part of their range, they also go out over the grassy plains. Wood frogs live in the southern Appalachian Mountains and north through the eastern and midwestern United States, Canada, and Alaska. No other amphibian in North America lives so far north.

In the eastern United States and southeastern Canada, another frog begins calling about the time the wood frog stops. This frog is the one that announces the beginning of spring, the spring peeper. The peeper is a tree frog so tiny it could sit on your thumbnail. It is heard much more often than seen. Its call is a high, clear note: *peereep, peereep, peereep, peereep.* It is repeated again and again, as long as the tiny frog is not frightened. If something startles the peeper, it stops calling.

Several peepers calling together have a different sound. It has been likened to the sound of sleigh bells softly tinkling in the distance. It can be heard for almost a half mile.

The peeper chorus is not just a bunch of frogs that happens to be calling together. Instead, the chorus is organized into

spring peeper

trios. The entire chorus can take a cue from any one member of any trio. A few moments after the first frog begins to call, the other two in its trio join in. Then, a few at a time, the other trios pick up the song. Soon all the peepers are calling. As they sing, their throats balloon into transparent bubbles, like glistening pearls.

The first mild, moist weather of spring brings out the peepers in force. They head for small bodies of water of all sorts. In the beginning the peepers call from bushes and shrubs near the water. After a few days they leave the brush and enter the water. They do not go far from shore but stay in the shallows. There they perch on fallen leaves, twigs, and stems. And they sing long into the night.

Shortly after the peepers arrive in their breeding waters, several other frogs and toads also begin to call. One of the earliest to be heard in the eastern United States and southeastern Canada is the American toad. All puffed up, with its vocal sac bulging, this warty toad may look foolish, but its call is as sweet as bird song.

The American toad makes a single, trembling note that some people think resembles the trill of a flute. Each call can last as long as a half minute. The toad repeats it over and over—often for hours at a time. By July the last American toads have finished breeding, and they are silent.

The call of the southern toad is similar to the American toad's. But it is higher and is repeated more rapidly. The

American toad

southern toad lives in the southeastern United States. Its call can be heard from March to October.

Not all toad calls are sweet and musical. The call of the Woodhouse's toad, for instance, is harsh and flat. The closest sound a person can make to it is to pinch his or her nose closed, and say *whaaah*. The note made by the Woodhouse's toad lasts just a few seconds, but it is loud enough to be heard almost a quarter mile away.

In the western states, the Woodhouse's toad begins to call as early as March. In the East, the toad starts to be heard in late spring. Its noisy calls rattle through the night all summer long. The Woodhouse's toad often goes by another name—Fowler's toad—in the eastern states.

A short, loud, rattly call is also the trademark of the gray tree frog. The noise made by a large chorus of these frogs can be almost deafening. Gray tree frogs begin calling in mid-spring. They live in the prairie states and east to the Atlantic Coast. Their choruses gather in brushy, wet meadows and shallow swamps and ponds. Gray tree frogs are only about two inches long. With bodies the color of bark, they are difficult to detect on the branches of trees.

The frog heard most often on the West Coast is another tree frog. This one, the Pacific tree frog, calls with a loud *kreck-ek*. Each call lasts only a second, but the frog repeats it in rapid-fire fashion.

Pacific tree frogs live in a vast region. They are found all the

gray tree frog

14

way from British Columbia in Canada to the tip of Baja California in Mexico, and as far east as Montana and Nevada. They are at home in high mountains, dry plains, and on islands off the Pacific Coast.

Within this broad region, the climate differs greatly. Therefore, the tree frogs breed at different times, depending on where they live. In the warm parts of this region, the frogs call and breed in January. Up north and in the mountains, they call in spring and early summer.

Leopard frogs also have a vast range. In fact, they have the greatest range of any frog or toad in North America. They live all the way from Canada to southern Mexico, except along the Pacific shore.

The leopard frogs of southern regions can be heard calling any month of the year. In places where it is very dry, the frogs breed whenever it rains enough to fill pools and puddles. That means they can be heard calling almost any time it pours. In most areas where leopard frogs are found, however, they call from March to May. The call is easy to describe—it sounds almost like a person snoring.

The call of the green frog is also easy to describe. It sounds exactly as though someone is plucking a loose banjo string. If you cannot imagine that, try this. Press your lips together, and then as hard as you can, say *plunk*.

Sometimes the green frog repeats this call three or more times in a row. Or else the frog calls, keeps quiet for several moments, and then calls again.

Green frogs, which live in the eastern states, call from late spring through most of the summer. Early in the breeding season, they call mostly at night. Later, they may call frequently in the afternoon as well. Like their calls, green frogs are big and powerful, with bodies more than three inches long.

The biggest frog in North America has the loudest call. It is the bullfrog, which is large enough to fill a soup bowl. Green as pond slime, the bullfrog calls with a deep, booming roar that seems to shake the night. Almost everyone who has heard it agrees it sounds like the words *jug o' rum* said very deeply. Bullfrogs, which live in every state but Alaska, call from late spring until the fall.

The mating calls of frogs and toads fill the night air from the time the snow melts to when the leaves fall. The songs of these amphibians mark the passing of the seasons. The clattering of the wood frogs tells of winter's end. The piping of spring peepers announces that spring has arrived. The booming of bullfrogs sings in the summer. And when the green frogs and bullfrogs stop calling, winter is not too far away.

One reason why bullfrogs and green frogs can be heard so late in the season is that they call even after they have finished breeding. Scientists have studied green frogs to learn why and think they have found out. The frogs sometimes call to keep other frogs away from their feeding territories.

Each green frog has its own feeding station. Unlike some other frogs, the green frog does not prowl about in search of

bullfrog

17

prey. Instead, the green frog sits in one place and waits for food to come to it. The owner of such a feeding territory warns other green frogs to keep away by calling every so often. This seems to keep the peace among the frogs.

Frogs make sounds for a number of other reasons too, but not often. Most frogs will open their mouths and scream loudly when attacked. Sometimes their screams startle the enemy so the frogs can escape.

Some kinds of frogs call in cloudy, humid weather, day or night. The large green tree frog of the southeastern states is one of these. Its call is like the clanking of a bell. Because it calls on cloudy, damp days followed by rain, people sometimes call it the "rain frog."

Another "rain frog" is the squirrel tree frog. When the weather is humid, this frog calls with a noise that sounds like the scolding of an angry squirrel.

Frogs and toads add great beauty to the night with their calls. They need very little in return. But what they must have to exist is water in which to breed. Where swamps and marshes are destroyed, and streams and lakes polluted, the frogs and toads sing no more in the darkness.

3 Night Birds

Hoots, hammers, and whistles

Night birds and the sounds they make in the dark seem spooky to many people. For sure, some of the calls of night birds may sound scary. But they are nothing to fear. Night birds call and sing for the same reasons birds do in the daytime. Sometimes it is to announce that the bird has claimed a nesting place, or that it is feeding, or to attract a mate.

The most familiar of all night birds are the owls. People usually describe the call of an owl as a hoot. But not all owls hoot. And even those that do may whistle, chirr, hiss, yelp, scream, wail, or rattle their bills.

Moreover, owls often sound different to almost everyone who hears them. If three or four people happen to hear the same owl at the same time, it still may sound different to each of them. Even experts on birds cannot agree on how to describe owl calls. Some scientists think, for example, that the call of the great horned owl sounds like this: *ho, hoohoo, hoo, hoo.* Others describe the call this way: *oot-too-hoo, hoo-hoo.* And still other bird experts say the call is *hoo, hoo-hoo-oo, hoot, hoot.*

Generally, though, each kind of owl seems to have a basic pattern to its main call. The number of hoots in the call usually remains the same or almost the same. The hoots in the call of the horned owl may number as few as three or as many as eight. However, the basic call is a series of five hoots. It is a deep, powerful call that travels a great distance.

The horned owl often makes its basic call while hunting. Every so often it stops its hunt, perches, and calls for several moments. Then it wings off again with barely a whisper of feathers. Many other owls have a similar habit.

The horned owl often follows the same hunting route, night after night. If the owl is heading in your direction, its calls will get louder each time it stops to perch. If the owl is going the opposite way, the calls gradually fade into the night.

A horned owl that is cornered behaves ferociously. It lifts its wings, thrusts out its head, and opens its beak. All the while it hisses like a steam engine.

During January and February, when the horned owls mate, they make sounds that are hair-raising. Perched near one another, the male and female owls cackle, hoot, yowl, and scream.

The horned owl gets its name from two tufts of feathers on

great horned owl

its head. It lives throughout most of North America. It is a huge bird with wings five feet across.

Sometimes people confuse the call of the horned owl with that of the barred owl. This big owl has a basic call of eight hoots. It is a more powerful call than the horned owl's but not as deep. Some people describe the barred owl's call as *hoohoo-hoohoo hoo hoo hoohooaw*. Others who have heard it say it sounds like *whoo who whoo who who to-hoo-ha*. From far away, it may sound like the barking of a dog.

The barred owl is likely to call from a perch in low, swampy woods. The horned owl usually calls from hillsides or dry woods. Horned owls hoot at almost any time of night. But barred owls call most frequently in the evening and just before dawn.

In some parts of the country the barred owl is called the "rain owl," because it often calls on overcast days. It is one of the few owls to commonly call before dark.

Often a barred owl will let loose with a burst of horrible screams, shrieks, and howls. No one knows why. But the calls of these owls have scared many people.

The owl heard by more people than any other is the little screech owl. It has a high, shivery call. Screech owls living in the eastern half of the United States generally have long, trembling whistles. Those of the western states make several short whistles, one after the other. Screech owls also click their bills and buzz loudly when frightened or excited. The reason people

screech owl

hear screech owls so often is that these small birds get along well near humans. Screech owls, smaller than pigeons, commonly live in backyards. They need only a hole in a tree or birdhouse to roost in. Their calls are heard in towns and cities as well as in the country.

More owls live in cities than many people suspect. Rats, mice, and pigeons provide plenty of food for city owls. Evergreens in parks make fine roosts. Belfries, abandoned buildings, and the undersides of bridges are excellent hiding places for owls, particularly barn owls. They have been discovered in the heart of cities, such as Washington, D.C., and New York.

Barn owls like to roost in old buildings. They are shy and come out only when the night is darkest. The heart-shaped face of the barn owl is ghostly white. No owl seems more spooky. It flutters and flits ever so lightly for a large bird. Its call is like the cry of a lost soul. The call is high and shrill—a wheezy hiss that screeches, rises, and falls.

The long-eared owl, with long feather tufts on its head, likes to roost in evergreens near open country, though it will sometimes stay in city parks, if they have stands of trees. This owl makes four different calls: It coos softly like a dove, mews like a cat, whistles, and hoots. Its whistle sounds like *whee-you*; its hoot is repeated several times in a row. This owl has several other calls too. It can yap like a puppy, squeal, and rattle its bill. The rattle means the owl is alarmed.

The little saw-whet owl also likes to roost in evergreens. This

barn owl

25

owl of the northern states never hoots, but it whistles endlessly. Its call is a short whistle—*too-too-too-too*—repeated faster than you can count. The owl may whistle a hundred times in a row without stopping.

Early in the spring when the saw-whet owl is breeding, it makes another call. This call is like the scraping sound of a metal saw being sharpened, or whetted. That is how the saw-whet owl got its name.

Several other night birds are also named after their calls. These birds include the whip-poor-will, poor-will, and chuck-will's-widow. They belong to a family known as the goat-suckers.

How this family got its name is an interesting story. A goatsucker called the nightjar lives in Europe. Like most other goatsuckers, it has a very wide mouth for catching insects on the wing. In the evening nightjars often chase insects flying near farm animals. The birds fly with mouths wide open. Long ago people noticed the open-mouthed nightjars flying near goats. The people imagined that the birds were trying to get near enough to the goats to drink their milk. The nightjars were nicknamed goatsuckers and the name stuck—not only to them but to their relatives around the world.

The calls of the whip-poor-will, chuck-will's-widow, and poor-will are clear, ringing whistles. The whip-poor-will calls the loudest and most rapidly of all. Sometimes it repeats its whistle two hundred times without stopping. The call of the chuck-will's-widow is similar but softer. The chuck is often very hard to hear. So is the third note in the poor-will's call, which really is like this: *poor-will-it.* The *it* can be heard only if the bird is very close.

The poor-wills live only in the far-western states. They like open country but not desert. The chuck-will's-widows inhabit woodlands of pine and oak in the southeastern part of the country. Whip-poor-wills are found in most of the region east of the Great Plains and in the Southwest. They sometimes call from city parks as well as in rural areas. In cities whip-poor-wills often perch on the roofs of flat-topped buildings.

The whip-poor-will, chuck-will's-widow, and poor-will call only while perched. Scientists believe the calls of these brown-colored birds mark nesting territory and, perhaps, attract mates. They never call in daylight or while flying.

However, another goatsucker does call before dark and on the wing. It is the nighthawk. Flocks of nighthawks come out before dusk and flit about after insects. As they fly, they call loudly. If you pinch your nose shut and say *peent*, you will have an idea of the sound made by nighthawks.

The calls of flying nighthawks easily reach the ears of people below. Flocks of nighthawks often appear over cities. They are after insects carried up by warm air rising from streets and buildings. The nighthawks continue to fly into the hours of darkness.

Nighthawks can be seen from coast to coast until the fall.

Then they gather in huge flocks and head for South America, where they spend the winter.

Several herons also call loudly in the night. The bittern, a kind of heron, makes a hollow, gurgling sound in its spring mating season. The call has been described in many ways. Here are some: *pump-er-lunk, oonck-a-tsiinck,* and *oong-ka-choonk.* You can see from this how difficult it is to translate some animal calls into words.

Only part of the bittern's call can be heard from a distance. Then it sounds like a sledge-hammer driving a wooden stake into the ground. This is the way people usually hear it. The bittern generally calls only from spots hidden deep within marshes.

Two other herons are well known for their calls in the darkness. The black-crowned night heron makes a loud, harsh *quowk.* The yellow-crowned night heron has a similar call in a higher pitch.

As they fly through the dusk toward feeding grounds, the herons call back and forth to one another. They probably keep in touch by their calls when it is too dark to see. The herons also call while feeding. Perhaps the sounds tell other herons where the food is. If the herons are frightened, they call very rapidly.

The black-crowned night heron lives throughout the United States but always near water. It is not fearful of man. Black-crowned night herons often can be seen in city parks near

black-crowned night heron

lakes, rivers, and the sea. During the day the herons roost in trees. As night approaches, they fly to the water's edge to feed. They eat fish, frogs, and other small water animals.

Yellow-crowned night herons are most common in the southern states. But a few of them live in the North as well. They feed in the daytime as well as after dark.

4 Insects After Dark

Chirps, buzzes, and zeets

The calls of insects are the most familiar sounds made by animals after dark. In most places, this is especially true during late summer and fall. Then countless millions of crickets and grasshoppers are calling. They call every night, all night long, as well as in the afternoon. Their calls make a steady drone. It seems never to stop. We can become so used to it in the background we can even forget it is there.

It also is easy to forget the real nature of the calls that make up the droning. They are signals made by the males to attract the females, which generally are silent. Some of the calls also mark the territory of the males.

Each species of cricket and grasshopper has its own set of calls, or songs. Often there are three different songs in a set. One brings the female near the male; another interests her in mating; the third is a rival song, used if another male approaches.

Two male grasshoppers sometimes wage "song battles" over territory. They face one another and sing the rival song until one tires. The male that keeps singing longest wins the territory—and any female that happens to be in it.

The grasshoppers noted for rival song battles are the short-horned species. They get their name because the antennae on their heads are rather short. The short-horned grasshoppers are the ones that make the drowsy, buzzing sound heard on summer afternoons and evenings. A common grasshopper in this group is the locust. It is the large one with black-and-yellow wings that flies up suddenly from the sides of roads and paths. The red-legged grasshopper is another member of this group. It is one of the grasshoppers most often seen in fields and pastures.

As the afternoon passes, large numbers of crickets add their calls to those of the short-horned grasshoppers. The field crickets and house crickets keep up a steady chirping. The chirping of the field crickets can be heard in the spring, but it really gets loud later in the summer. It is one of the most familiar sounds in nature. Actually, each *chirrup* is made up of several sounds, but our ears cannot separate them.

If a male field cricket succeeds in attracting a female, he dances about and changes his song. The new one is very high pitched, so high we can barely hear it. The male house cricket also changes his tune if a female approaches. But this song can be heard. It is a loud trill.

short-horned grasshopper

The soft but throbbing trills heard in the background of the insect orchestra are made by the ground crickets. Their song goes on and on and on. It is higher pitched than the calling of the field and house crickets.

Cricket calls increase as darkness approaches. After dark the loudest of all crickets join the insect chorus. They are the tree crickets. Ghostlike creatures, tree crickets are pale and slim, with long, transparent wings. They hide in the shrubbery and trees. From midsummer to late fall they sing the night away.

One of the most common of all night sounds is made by the broad-winged tree crickets. The call of these insects is a high, sweet trill lasting several seconds. The crickets repeat it again and again. During the last weeks of summer these crickets perch almost everywhere in branches. Their calls seem to come from every direction.

The narrow-winged tree cricket has a purring call. Each purr lasts two seconds. After two more seconds, of silence, the cricket repeats its call again.

The snowy tree cricket calls for hours on end with a high, steady *treeet, treeet, treeet.* This cricket calls all night long, stopping only with the dawn.

Night is also the time when the long-horned grasshoppers start calling from branches. These insects have antennae that are as long as their bodies. And their bodies may be two or more inches in length.

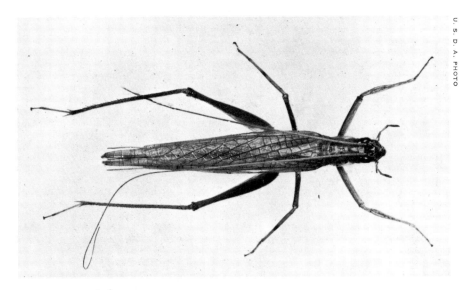

tree cricket

The best-known of the long-horned grasshoppers is named after its call. This insect is the katydid. The common katydid of the eastern United States is the one that buzzes *Katy did, no she didn't* over and over in the darkness. The more wide-spread fork-tailed bush katydid has only one note in its call. It is a high-pitched *zeet*. This call is often lost amidst the songs of other insects, because it carries only a short distance. The common eastern katydid, however, can be heard a mile away.

The big angular-winged katydid also has a loud call. It goes like this: *tzeet, tzeet, tzeet, tzek, tzz*. The last notes are made more slowly than the first ones in the series. Another, smaller kind of angular-winged katydid calls with a loud, sharp *zeet, zeet, zeet, zeet*.

35

katydid

The calls of the long-horned grasshoppers and crickets really are the noise of the insects' wing covers rubbing together. On their wing covers, these insects have structures called a file and a scraper. They are made of chitin, the same tough, light material that covers the rest of the insect's body. The file is rough, covered with notches, or teeth. The scraper has a hard, clean edge.

The common katydid has a file under its left wing cover and scraper atop its right. They are located just behind the insect's shoulders. When the katydid calls, it raises its wing covers, with left overlapping the right. Then it moves the wing covers, and scraper and file grate together, making a noise.

Near the scraper and file are two small "drumheads" of chitin. They are thinner than paper. Yet steel of the same thickness would not be as strong. As the disks vibrate to the noise made by the file and scraper, the noise gets louder. We hear this as the call of the katydid.

Crickets call in a similar manner. The male field cricket uses a file under its right wing cover and a scraper atop the left. The cricket places its right wing cover over the left. Then it moves its wings up and down. Scientists believe the noise of the scraper and file is made on the downstroke. Like the katydid, the cricket has sound disks which boost the volume of its call.

Most short-horned grasshoppers rub scrapers on their legs against the wing covers. The wing covers vibrate and produce sound. Some of these grasshoppers call just by rubbing their wing covers together.

The females of most crickets and grasshoppers receive the calls of the males with ears surprisingly like our own. However, their ears are in very different places than ours. Short-horned grasshoppers have an ear on either side of the body. Long-horned grasshoppers and crickets have one just below the "knee" of each foreleg.

Once a female hears a male's song, she follows it to him. After she mates with the male, her hearing no longer is important. The female short-horned grasshopper actually becomes deaf after mating. The fertilized eggs in her abdomen grow and press upon her ears until she loses her hearing.

Female tree crickets, unlike their relatives, never hear the songs of the males. The male tree crickets sing only for one another, to mark territory. But even so, when they call, they attract the deaf females. How? Not by sound but by scent.

When the male raises his wings to call, a tiny bowl is uncovered on his back. The bowl is filled with a thick fluid. The scent of the fluid draws the female, which wants to drink it. To reach the fluid, the female must climb up on the male's back. When she does, the male mates with her.

The calls of some insects can be used to figure out the temperature of the air. Here is one way: Count the number of times a snowy tree cricket calls in fifteen seconds. Add the number *39* to the total number of calls. The sum is the temperature in degrees Fahrenheit.

Another way to do it is to write this formula:

If N is the number of chirps per minute, and
T is the temperature,

$$T = 50 + \frac{N-92}{4.7}$$

If you can figure formulas, here is another one, using the house cricket:

$$T = 50 + \frac{N-40}{4}$$

One scientist counted the number of times the common katydid calls at different temperatures. He found that at a temperature of 82 degrees Fahrenheit, the insect calls eighty-nine times a minute. At 68 degrees, the katydid calls thirty-eight times a minute. It calls only twenty times a minute when the temperature is 58 degrees. Katydids stop calling entirely below 50 degrees. Field crickets stop calling at a temperature of less than 55 degrees.

The last insects to call in the fall at night are the tree crickets. They sometimes can still be heard after the first heavy frost, if the weather warms up again. But as winter approaches, even the tree crickets drop into silence. Before the snow flies, the insect musicians have died. However, their eggs or young remain. They survive the winter. In the spring they grow into adults. By midsummer they are singing in the darkness.

5 Mammals of the Night

Howls, yowls, and yips

A great number of mammals are active at night. As a group, in fact, the mammals are mostly night animals. People, apes, and monkeys are among the few which are not. However, even though mammals may be all around us in the dark, we seldom hear them.

Night mammals are usually very quiet. The fox trots through the woods as silently as a blown leaf. On tiny white feet the deer mouse slips like a whisper through the grass. The muskrat swims with scarcely a ripple.

Sometimes, though, even the most silent of mammals make lots of noise. When alarmed, mice squeak rapidly and loudly. We cannot hear them, for their calls are too high pitched, but they alert other mice to danger.

Muskrats often feed in late afternoon but also come out at night. When a muskrat is alarmed, it dives below the surface with a loud splash of its tail. People who hear the splash in the dark sometimes mistake it for the sound of a big fish jumping. The muskrat's relative the beaver makes a famous alarm signal with its tail. When the beaver is frightened, it smacks its broad, heavy tail on the water with a thunderous *whack.* In the quiet of a wilderness night the noise can sound like an explosion.

Rabbits that are attacked make an even more startling noise. It is a shrill, terrible scream, amazingly loud. The scream of a jackrabbit can be heard a mile away. The snowshoe rabbit and cottontail rabbit also scream horribly when in trouble.

A raccoon that is alarmed screams loudly, too. The scream is a version of the raccoon's basic call, which is a churring sound.

When a mother and her young raccoons are feeding in the dark, they keep in touch by churring softly. If danger threatens, the mother warns the young by churring loudly. The warning churr is high and shrill. If the danger increases, it rises to a scream. Such a sound in the night can mean that a great horned owl, dog, or other enemy is after the raccoon's young. If the raccoon is fighting with an enemy, the coon may sound like a dozen different animals, screaming, screeching, and spitting.

People sometimes mistake the screeches of an angry raccoon for the calls of a bobcat. These wild cats, however, are usually silent. But in their mating season they can be very noisy. The mating season of bobcats is in late winter. Both the male and

raccoon

cottontail rabbit

bobcat

female call then. They sound just like yowling domestic cats, only much louder.

Male and female mountain lions also make wild, yowling sounds when mating. At other times mountain lions let loose with weird, hair-raising screams. Some people believe they sound like the screams of a terrified woman. Because of this call, frontiersmen called the mountain lion the "screamer."

During the middle of winter the barking of the gray fox and the red fox can be heard. Midwinter is the mating season for both species. The gray fox has a rough, gravelly bark. It also barks frequently when it is teaching its young to hunt, late in the summer. The red fox has a sharp-sounding bark. It also makes a yapping call when it senses someone approaching its den.

No wild mammal in North America has a call as well known as the coyote. Its howling, yipping song is familiar even to people who never have seen the creature. However, much about the coyote's song remains mysterious. Scientists do know that coyotes usually sing in family groups. The singing probably marks the territory of the family. It also is known that coyotes sing most during their mating season, in January and February. Scientists suspect there are many other reasons why coyotes sing, but no one knows what they are.

One of the reasons a pack of wolves howls may be to claim its territory. Wolves belonging to other packs probably hear the howling and stay away. Howling also seems to keep a wolf

mountain lion

red fox

NEW YORK ZOOLOGICAL SOCIETY PHOTO

wolves

pack together when it hunts at night. The wolf pack usually has a dozen or more members. While hunting, the wolves cover many miles of territory. The wolves may keep in touch by sound rather than sight.

When several wolves are kept in a large zoo exhibit, they will howl at night just as they might in the wild. A single wolf usually begins the howling. One by one the others quickly add their voices. Some stay seated on their haunches. Others stand. Within a few moments the entire group has gathered together, noses pointed to the sky. Their wild music echoes over the landscape. It will raise goose bumps on anyone who hears it.

Unless you happen to live near a zoo, there are few places in the United States where you can hear wolves howling in the night. Alaska still has many wolves. But south of Canada the howling of the wolf pack can be heard only in the upper Great Lakes region and a few parts of the Rocky Mountains. The wolf is a creature of the true wilderness. When the wilderness disappears, so does the wolf. Moreover, people have persecuted the wolf without mercy. Even today, in many places where wolves remain, they are killed on sight.

The disappearance of the wolf has helped its cousin the coyote. Coyotes have spread into much of the territory once roamed by wolves. Coyotes now live as far east as New England. In the western states, coyotes can be found on the edges of large cities.

It may be surprising how many other mammals mentioned

here sometimes turn up in cities. Foxes often do. Raccoons and muskrats are common city residents. Cottontail rabbits can live right in backyards, as long as they can find grass, weeds, vegetables, or flowers to eat. Even mountain lions sometimes are found near very large cities. These big cats have been seen in the outlying parts of Los Angeles. A few of them live not far from Miami, Florida. And recently there have been reports of mountain lions seen only forty or fifty miles from New York City.

6 A Time to Listen

Few people are lucky enough to hear the screams of a mountain lion or the howling of a wolf pack in the wilderness night. But most of us can hear many other creatures that are awake after dark. They are all around us, in the country, town, and city. And they can be just as fascinating as wolves, mountain lions, or other big wilderness creatures—if we understand their ways.

The trouble is, many people who hear these night creatures never guess what is making the noise. All these people know is that something is making a funny sound in the darkness. What and why remain mysterious. Yet if these people understood the meaning of the sound, they might be surprised, even excited. The night would have more magic, and they would learn more about their world.

The world need not be mysterious at night. In fact, night is a time when we can perceive our world more in new ways. Our eyes may not work so well after the sun goes down. But our ears can serve us very well. Night is the time to listen to our world. Try it tonight.

For Further Reading

Cohen, Daniel. *Night Animals.* New York: Julian Messner, 1970.

Evans, William F. *Communication in the Animal World.* New York: Thomas Y. Crowell Company, Inc., 1968.

Milne, Lorus J. and Margery J. *The World of Night.* New York: Harper & Row, Publishers, Inc., 1956.

Tinbergen, Niko. *Animal Behavior.* (Life Nature Library: Young Readers Ed.). Alexandria, Va.: Time-Life Books, 1968.